# What Is Weather?

Nellie Wilder

The weather is
sunny.

I am warm.

The weather is
windy.

I am **chilly.**

The weather is rainy.

I am wet.

The weather is
snowy.

I am cold.

The weather is
**humid.**

I am hot.

The weather is
cloudy.

I am cool.

The weather is foggy.

I am **damp**.

The weather is
stormy.

I am snug in my home!

# Let's Do Science!

What kinds of weather can you observe? Try this!

## What to Get

❏ paper and pencil

# What to Do

**1** Make a chart like this one.

| Day | Weather |
|---|---|
| Monday | |
| Tuesday | |
| Wednesday | |
| Thursday | |
| Friday | |

**1**

**2** Go outside. What is the weather today? Write or draw the weather in your chart.

**2**

| Day | Weather |
|---|---|
| Monday | rainy |
| Tuesday | |
| Thursday | |
| Wednesday | |
| Friday | |

**3** Do this every day for two weeks. Do you notice any patterns? What do you predict tomorrow's weather will be?

**3**

| Day | Weather |
|---|---|
| Monday | rainy |
| Tuesday | sunny |
| Wednesday | cloudy |
| Thursday | |
| Friday | |

# Glossary

**chilly**—cold

**damp**—lightly wet

**humid**—warm and damp

# Index

# Your Turn!

Go outside. Use all your senses to describe the weather.

## Consultants

**Sally Creel, Ed.D.**
Curriculum Consultant

**Leann Iacuone, M.A.T., NBCT, ATC**
Riverside Unified School District

**Jill Tobin**
California Teacher of the Year
Semi-Finalist
Burbank Unified School District

### Publishing Credits

Conni Medina, M.A.Ed., *Managing Editor*

Lee Aucoin, *Creative Director*

Diana Kenney, M.A.Ed., NBCT, *Senior Editor*

Lynette Tanner, *Editor*

Lexa Hoang, *Designer*

Hillary Dunlap, *Photo Editor*

Rachelle Cracchiolo, M.S.Ed., *Publisher*

**Image Credits:** Cover & pp.1, 3, 6, 8, 14 iStock;
p.17 Blend Images/Alamy; p.15 LWA/Getty Images;
pp.18–19 (illustrations) Rusty Kinnunen; all other images
from Shutterstock.

**Library of Congress Cataloging-in-Publication Data**

Wilder, Nellie, author.
  What is the weather? / Nellie Wilder.
    pages cm
  Summary: "It is time to learn about kinds of weather."—
Provided by publisher.
  Audience: K to grade 3.
  Includes index.
  ISBN 978-1-4807-4530-8 (pbk.) —
  ISBN 978-1-4807-5139-2 (ebook)
1. Weather—Juvenile literature.
2. Readers (Primary)  I. Title.
  QC981.3.W536 2015
  551.5—dc23
                              2014008932

### Teacher Created Materials

5301 Oceanus Drive
Huntington Beach, CA  92649-1030
http://www.tcmpub.com
**ISBN 978-1-4807-4530-8**